Teddy stretched and yawned. It was the morning of the big day. He was excited because this was the day his family was going to visit the **Space Center**. He was going to see **rockets** and the people who build them. He had so many questions. He wanted to learn all about space, and this was going to be fun.

"Slow down, Teddy! You are eating too fast." Said Teddy's Mom. Mrs. Tusker had made breakfast for the family before their big trip. "I'm so excited, Mom. I can't wait to go to the **Space Center**." "Listen to your Mother, Teddy." Mr. Tusker said as he looked over his cup of coffee.

"Yeah, listen to mom Teddy." His sister Tanya winked at Teddy. Teddy winked at his little sister. Tanya was four years old and carried her Baby Bear with her everywhere.

Alright, mom. I made my bed and brushed my teeth. Is it time to go to the **Space Center** now?" Teddy was running out of his room and throwing on a coat. "Yes, Teddy, we are all ready." His mom was taking Tanya by the hand to go outside. "All aboard the Tusker Train! Next stop the **Space Center**." Mr. Tusker was pretending to be a train conductor and laughing as he went outside.

"Wow! There are some big buildings at the **Space Center**. I can see them from here." The big buildings looked very tall to Teddy, and the sunshine was making them very bright and shiny. "That's right, Teddy." Mr. Tusker said.
"They have to be very large as the **rockets** are very tall, and that is where they make the rockets before they are **launched** into space."

"Let's go there, first, Dad!" Teddy was pointing to a big building with an American Flag painted on it and pulling his dad in that direction. "It is so tall. There must be a **rocket** in there." Mr. Tusker was happy; Teddy was so excited and wanted to learn about space. "Yes, Teddy. The pieces or sections of the rocket are delivered to the **Space Center**, and then they put it together inside that building. The building keeps the rocket safe from rain and wind while they put the pieces together."

"Wow! Dad, there is a **rocket** here. It is so big and tall." Mr. Tusker smiled and says, "It has to be very tall, Teddy. You see, to get off the ground and up into space takes a lot of **power**. The **rocket** is made up of sections. Each section helps the **rocket** get into space, and then it falls away when it is no longer needed. When that section comes off, it helps lighten the **rocket** for its travel into space."

"That's right, Teddy." A man in blue had walked over and was bending down to shake Teddy's hand. "I'm Andy, the **Astronaut**, and I am glad to meet you." Teddy looked up at Andy with his eyes wide in surprise.

He put out his hand, and the two shook hands. "Wow! You are an **Astronaut**! I'm Teddy Tusker, and I am glad to meet you. I want to learn all about **rockets** and space" Andy smiled and stood up straight. "Very pleased to meet you, Teddy."

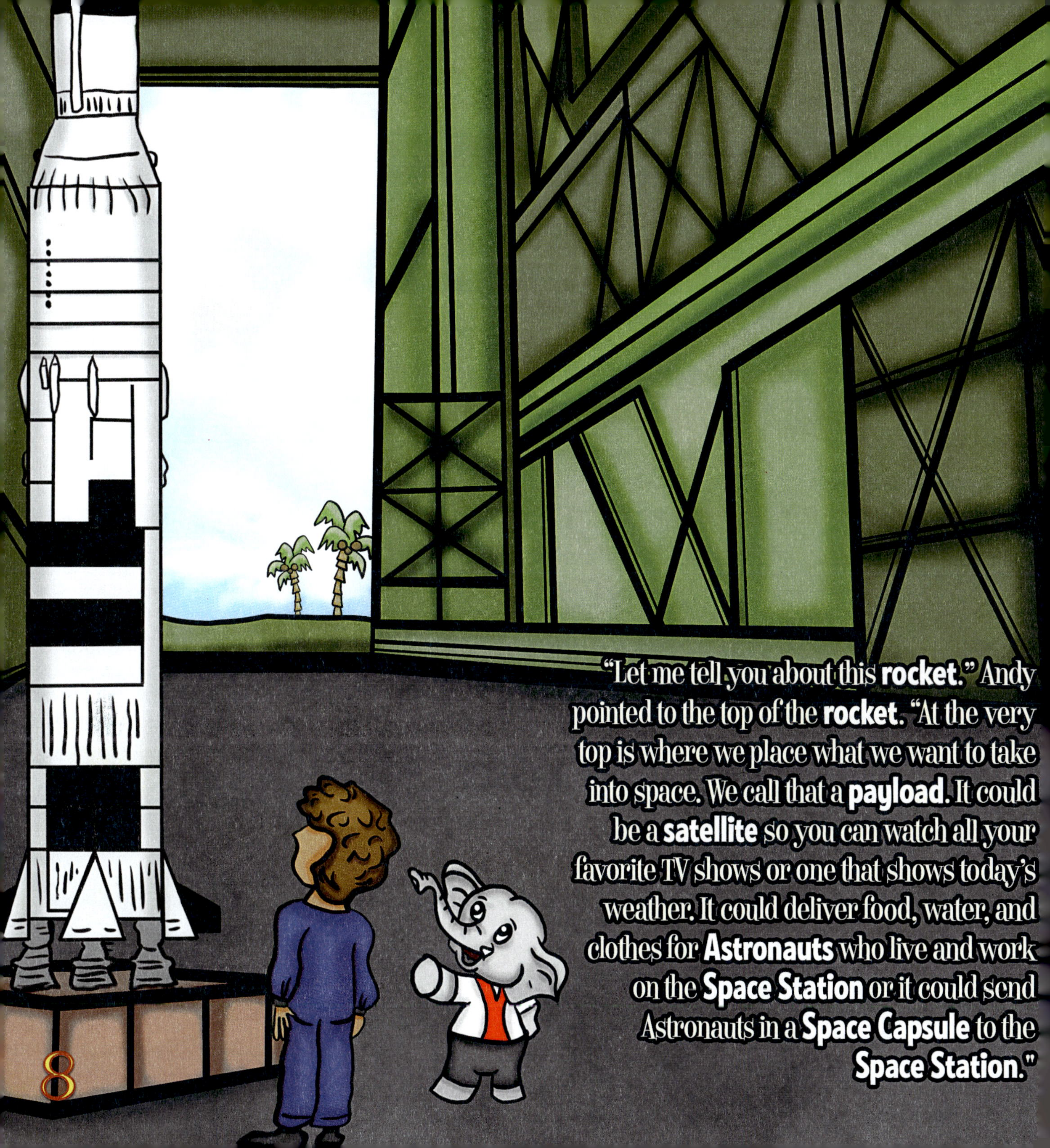

"Let me tell you about this **rocket**." Andy pointed to the top of the **rocket**. "At the very top is where we place what we want to take into space. We call that a **payload**. It could be a **satellite** so you can watch all your favorite TV shows or one that shows today's weather. It could deliver food, water, and clothes for **Astronauts** who live and work on the **Space Station** or it could send Astronauts in a **Space Capsule** to the **Space Station**."

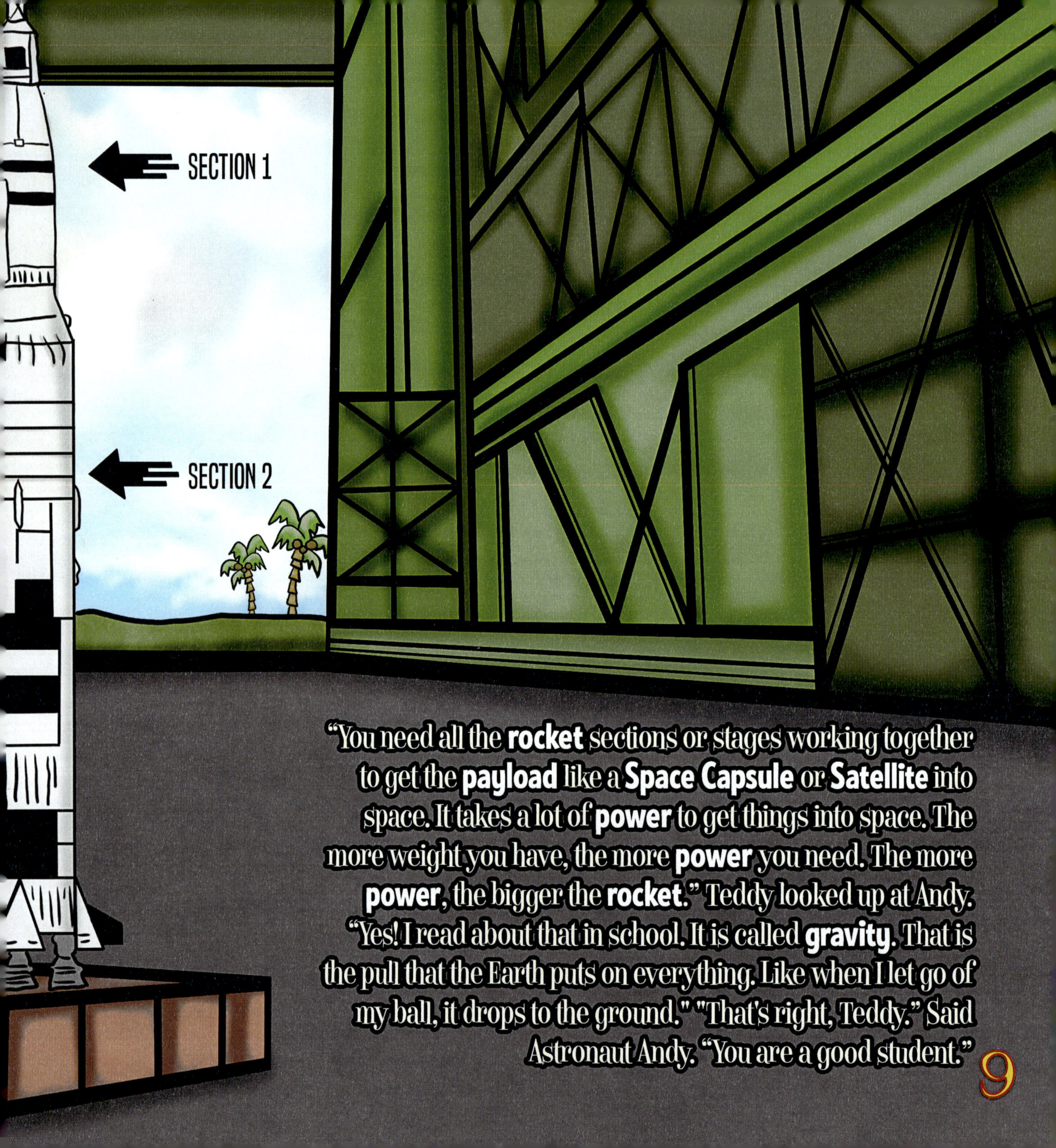

"You need all the **rocket** sections or stages working together to get the **payload** like a **Space Capsule** or **Satellite** into space. It takes a lot of **power** to get things into space. The more weight you have, the more **power** you need. The more **power**, the bigger the **rocket**." Teddy looked up at Andy. "Yes! I read about that in school. It is called **gravity**. That is the pull that the Earth puts on everything. Like when I let go of my ball, it drops to the ground." "That's right, Teddy." Said Astronaut Andy. "You are a good student."

"So, the **rocket** sections are put together in this building. Where does it go to **launch** into space?" Teddy asked. "Well, the workers put it on a large moving **platform** called a "**crawler**". They build the rocket on that, and then they move it out of the tall building to a **Launch Pad**."

Astronaut Andy pointed off in the distance as they walked out of the tall building. "Oh, that." Said Teddy. "I wondered what the tall structure was. Can we go see it?" "Of course." Said **Astronaut** Andy.

"Wow. This is a tall building too. I guess it has to be for the **rocket**?" "You are right, Teddy." Andy was looking up. The **rocket** is delivered to the **Launch Pad** after being assembled. It is connected to this **structure** to help keep it from blowing over in the wind and to get all the **commands** it needs before blasting off into space.

"**Commands** like when to start the **engines**?" Asked Teddy. "You are right again, Teddy." Andy smiled.

"Where do the **commands** to **launch** the rocket come from?" Teddy asked. "Well, Teddy, that will be the next stop on our tour. The building close to where they build the **rockets** is where the **commands** go to the **rocket**. This **rocket** is going to the **Space Station** soon. It will deliver food, clothes, and **experiments** to the **Astronauts** living up there in space." said Andy pointing up to the sky.

"This building is called the **Launch Control Center.** People who support the launch sit in this building. They send **commands** to the **rocket.** They also **check** the **rocket** to make sure it is ready to **launch** into space." **Astronaut** Andy was walking towards the long building and talking to Teddy and his family.

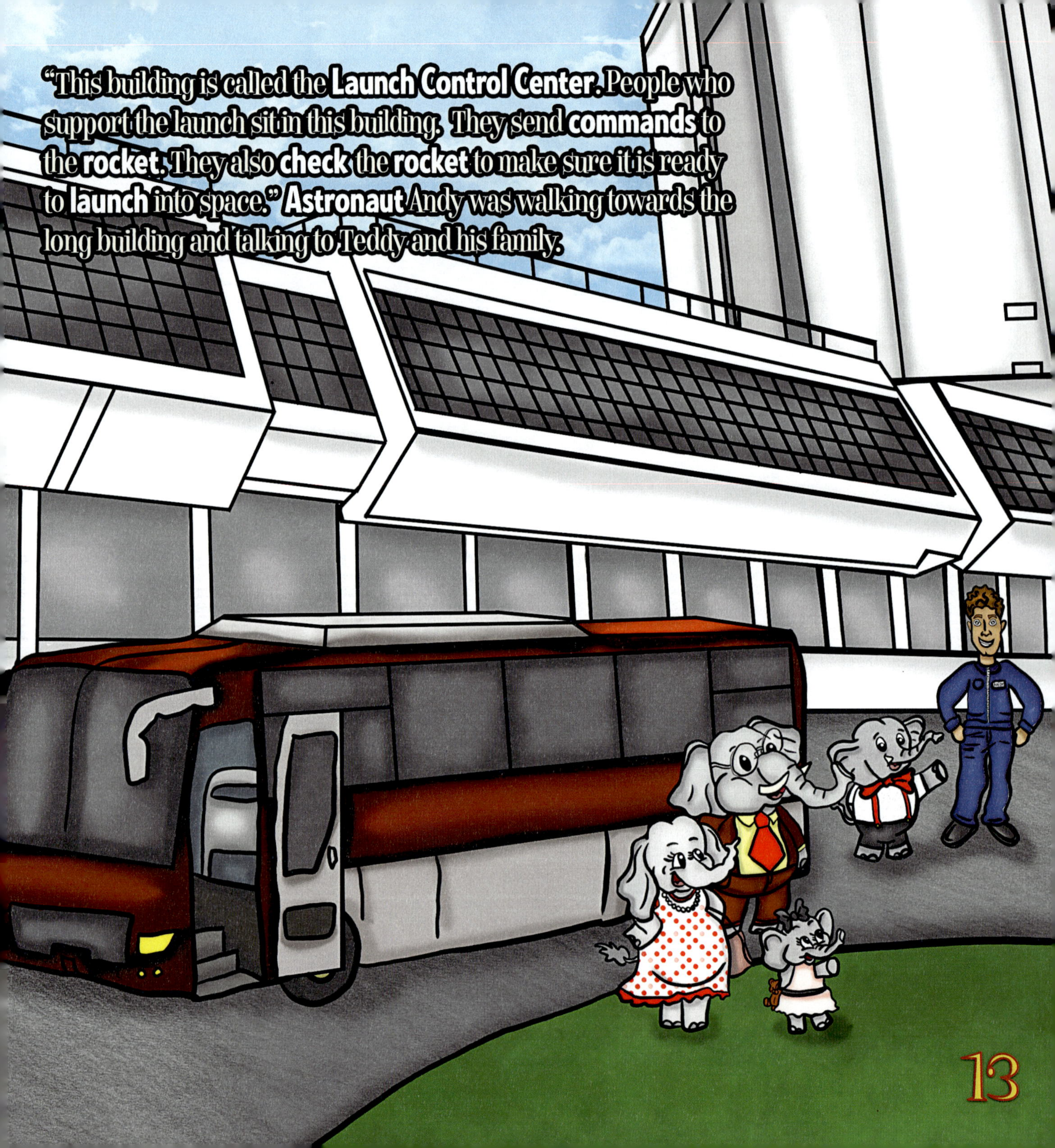

"All of these **computer screens** will have a **launch** team person sitting at it the day of the big **launch**." **Astronaut** Andy was pointing to all the empty **workstations**. Teddy's eyes got wide as he said, "I know this one from school. All the **engineers** and **technicians** sit at these **workstations** and make sure the **rocket** is ready. They have different jobs like the **engines**, **navigation**, and **fuel**, right?" "That's right, Teddy." Andy said as he smiled and winked at Teddy.

"Ok. I think I understand how the **rockets** are built and **launched**. What goes inside them to go into space?" Teddy was looking up at Andy as they walked to the visitor bus. Andy answered, "That will be next. Teddy, I think you would like to see where we get **Astronauts** ready to board the rocket. As I said before, the **rockets** carry **Astronauts**, or they carry **equipment** or **satellites**. Some even carry **vehicles** that go to other planets."

"Wow. Go to other **planets**?" Teddy looked excited. "That's right, Teddy. We have a **robot rover** on **Mars** right now. The **rovers** go where the **environment** is too harsh for humans.

They **explore** and send back data on what it is like on the planet. That **information** will help us decide if we want to send humans there."

"Humans? On **Mars**? Really?" Teddy looked surprised. "Yes, Teddy, someday we will send humans to visit **Mars**."

"In this building, we have a place where the **Astronauts** stay before they get on a rocket. They come here before **launch** and practice what they will do in space. If you want to be good at something, you have to practice a lot." "Yes, sir. That is what my baseball coach says." "He's right, Teddy. I am glad you practice." Andy was leading the Tusker family into the building.

"Here is the Suit Room. This room is where all the Astronauts get into their **Space Suits**. These are the suits they will use when riding in the **rocket**. These are not the suits that they wear in space outside the **rocket**. Those are called **Extra Vehicular Mobility Units** or EMUs." Teddy looked up at Andy and said, "I know about the EMUs. They keep the **Astronauts** safe and comfortable in space."

Teddy began to daydream as Andy talked.

"That's right, Teddy. Space is a very harsh **environment** for you and me. There is no air to breathe, and it is cold. We could not **survive** in space without an **EMU. Astronauts** must wear them when they go outside in space to do work. It could be to repair a piece of **equipment** or something else important they need to do outside the **Space Capsule** or **Space Station.**"

"Someday, I hope I can be an **astronaut** like you, Andy. I would like to ride a **rocket** in a **space capsule** and go to the **Space Station**." Andy looked at Teddy, "You can, Teddy. It is important to learn about math and science. Do you like math and science, Teddy?"

"Oh, my yes, those are my favorite subjects in school." Teddy said. Andy looked at Teddy and smiled. "You could be an **astronaut** too, Teddy. Study hard."

"Thank you so much for being our guide on the tour of the **Space Center**, Mr. Andy. I learned a lot and look forward to the next **launch** of a **rocket**." "It was my pleasure," said **Astronaut** Andy. Mr. Tusker, Mrs. Tusker, Tanya, and Teddy all waved good-bye to **Astronaut** Andy and climbed into the car to go home.

"It's almost time for the **rocket** to blast off!" Teddy yelled as he ran outside the Tusker home. It had been a week since visiting the **Space Center**, and today was the day of the big launch. "Yes, Teddy, it should blast off any moment." Mr. Tusker said as he adjusted his camera. "You can see it from here, Teddy." Mrs. Tusker said, and Teddy ran up beside her. "Is it going to be loud, Teddy? Tanya asked. "Boy, I hope so." Teddy answered.

Just then, a bright light appeared over the hill in front of their house, and, slowly at first, the **rocket** began to climb into the sky. The **rocket** roared into the air on a column of smoke and fire. A massive roaring sound of the **engines** came rushing towards them. The rocket continued to climb up and up until it became a just a dot of light and then disappeared. The roaring sound faded away to silence.

As Teddy went to sleep that night, he began to dream of **rockets**, the **Space Center** and **Astronaut** Andy. He knew someday he too would be an **astronaut** and ride on a **rocket** to the **Space Station** with **Astronaut** Andy.

VOCABULARY

WORD	MEANING/DEFINITION	USED IN A SENTENCE
Astronaut	A person trained to take part in space flight.	The astronaut climbed on board his spaceship.
Commands	To order or instruct to do something.	The computer sent commands to the printer to print a page.
Computer	An electronic device that is used to store and sort information and work with data at a high speed.	I like to play games on my computer, but I also use it to study. The library has a lot of computers that anybody can use.
Computer Screen	A television like device that shows pictures, movies, games that are attached to a computer	My computer screen helped me do my homework by showing me the internet on my computer.
Data	Facts, figures, or other pieces of information that can be used to learn about something.	The computer used its data to make a new discovery.
Energy	The power or ability to make something work or be active.	Plants use the energy of the sun to grow
Engine	A machine that uses energy from fuel or electricity to do work, such as to move.	The car engine made the wheels turn, and the car move.

VOCABULARY

WORD	MEANING/DEFINITION	USED IN A SENTENC
Engineer	A person who is trained in the use or design of machines or engines, or other technologies.	I have a friend who is an electrical engineer.
Equipment	Things made or used for an activity.	The coach had us gather up our baseball equipment after practice.
Experiment	A carefully planned test used to discover something unknown.	We put water on one plant and vinegar on another as an experiment to see if water wa better for plants.
Extra Mobility Unit	A suit that is worn by astronauts to protect them from the harsh environment of space.	The astronaut wore his Extra Mobility Unit so he could fix the broken rocket while floating in space.
Fuel	Anything such as wood or gasoline that is burned as a source of energy.	A car cannot run without fuel
Gravity	The force by which all objects in the universe are attracted to each other.	On Earth, objects fall to the ground because of gravity's pull.

VOCABULARY

WORD	MEANING/DEFINITION	USED IN A SENTENCE
Information	Knowledge or facts that come from a source.	The newspaper is full of information.
Launch	To put in motion with force.	We are waiting for the rocket to launch into outer space.
Launch Control Center	A building that manages the countdown of a rocket or spaceship.	The Launch Control Center gave the commands for the spaceship to launch.
Space Suits	A suit worn by astronauts when they are launched into space. It helps protect them on the ride up into space.	The astronauts put on their space suits before entering the spaceship.
Launch Pad	A structure where rockets are sent into space	The rocket stood on the launch pad, ready to blast off.
Mars	The seventh-largest planet in the solar system and fourth in the distance from the sun. Mars is known as "the red planet" because of its unusual	Mars is called the "Red Planet" because it looks red when looking at it from Earth.
Monitor	A screen for a computer that shows information. Also, a "Computer Screen."	My monitor allowed me to see what was happening in my video game.

VOCABULARY

WORD	Meaning/Definition	USED IN A SENTEN
Navigation	The act of setting a course for or controlling a ship or aircraft.	The ship used a navigation map to get from the island to the port in my town.
Platform	A raised level surface used as a stage or area in which to perform work.	The band got on the raised platform so everyone could see them when they played their songs.
Power	Energy that can do work. (See Energy)	The rocket had power form i engines to roar into space.
Robot	A machine that can perform some of the same tasks as a human being.	Robots do some of the work building cars.
Rocket	A flying device shaped like a tube that is driven by hot gases released from engines in its rear. Rockets are used to launch fireworks, signals,	Many rockets are launched from the Kennedy Space Center in Florida.
Rover	A type of vehicle designed to travel on the surface of a planet, comet, or moon, to provide transportation or retrieve scientific data.	The rover on the surface of Mars is named "Opportunity."

VOCABULARY

WORD	MEANING/DEFINITION	USED IN A SENTENCE
Satellite	A spacecraft that is sent into orbit around a planet or other object in space to gather or send back information to Earth.	The satellite above the Earth gave us pictures of the weather in our town.
Space Capsule	A spacecraft designed to carry humans into space.	The astronauts climbed into their space capsule before blasting off.
Space Center	A large facility with many buildings that are used to launch rockets and spacecraft into space.	All the rockets that we launch are at the Space Center.
Space Station	A facility in space that is orbiting a planet or moon that is used to house human beings in space.	The rocket delivered the new astronauts to the Space Station.
Structure	Anything that has been built for a purpose.	The new structure at the playground will be used as a "Jungle Gym."
Survive	To continue to live despite a serious threat to one's life.	They had so little food that it was a miracle they could survive without it.

VOCABULARY

WORD	MEANING/DEFINITION	USED IN A SENTEN
Technician	A person whose work requires special skills	To be an electrical technicia my father had to go to colleg
Vehicle	Something used to carry and move people or things.	Cars, buses, and airplanes ar vehicles that carry people around the country.
Workstation	A work area that is usually designed for one person and is equipped with the necessary tools or materials to perform a particular job.	The office where I work has ten workstations so people can concentrate on getting their work done.

SPECIAL SHOUT OUT

I WOULD LIKE TO ACKNOWLEDGE THE HELP AND ENCOURAGEMENT FROM RELATIVES AND FRIENDS THAT I RECEIVED DURING THIS PROJECT. KENT, DALE, LAURIE, BRIGITTE, ROXANNE, AND SUSAN. YA'LL ROCK. AND COURSE MY LORD AND SAVIOR BECAUSE WITHOUT HIM I AM NOTHING.